Predicting the

TUESDAY   WEDNESDAY

83   83   85

by Katherine Scraper

I need to know these words.

chart

clouds

rain

snow

tools

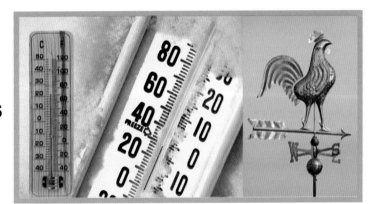

wind

We can watch the weather. We can look at **clouds**. We can see the **wind** blow.

▲ Which flag shows that the wind is blowing?

▲ Some clouds bring rain.

We can see **rain**. We can feel how hot or cold the air is.

▲ Do you like hot days?

▲ Do you like cold days?

▲ Do you like rainy days?

We can measure weather, too.
We use **tools** to measure
the weather. We can measure
the temperature.

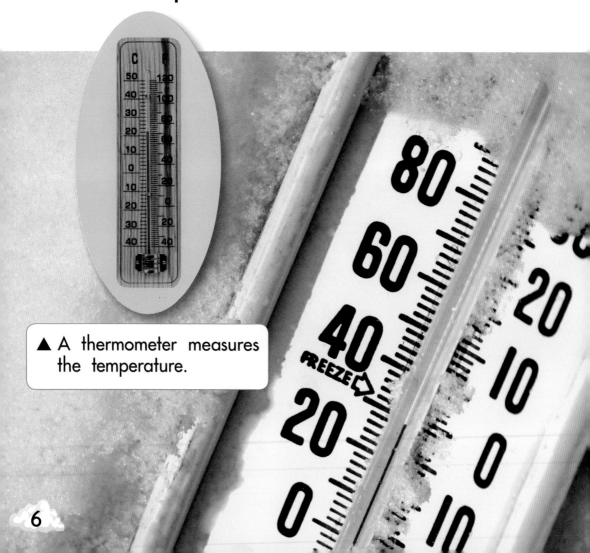

▲ A thermometer measures
the temperature.

We can measure which way
the wind blows.

A weather vane measures ▶
which way the wind blows.

We can record information about the weather. The things we record are data. We can put the data on a **chart**. Then we can look at all the data together.

How is the
weather today?

sunny	
windy	
rainy	
cloudy	
foggy	
hot	
cold	

▲ Clouds, wind, and temperature
are all parts of the weather.

9

We can use data to predict the weather. The data helps us predict the weather each hour.

Time	Condition
9:00	Sunny **75** degrees
10:00	Partly Cloudy **74** degrees
11:00	Rain **72** degrees
12:00	Thunderstorms **72** degrees
1:00	Partly Cloudy **73** degrees

▲ What will the weather be like at 10:00?

The data helps people predict the weather each day.

Sunday	Monday	Tuesday	Wednesday	Thursday	Friday	Saturday
Mostly Cloudy	Rain	Thunderstorms	Sunny	Sunny	Partly Cloudy	Rain
High 51° Low 38°	High 49° Low 39°	High 50° Low 42°	High 56° Low 47°	High 57° Low 45°	High 55° Low 41°	High 54° Low 41°

▲ What will the weather be like on Wednesday?

We can look at some data.
This data shows clouds in the air.
The clouds can bring rain
or **snow**. The wind is blowing
from the north. The temperature
is very cold.

	What do the clouds look like?	Which way is the wind blowing?	What is the temperature?
Sunday		from the north	30 degrees Fahrenheit −1 degree Celsius

▲ Rain clouds and snow clouds can look alike.

We can predict snow.

▲ These clouds brought snow.

Why do we predict the weather? What do we want to know? We want to know what clothes to wear. We want to know what things we can do.

◄ What do you wear on a hot day? What do you wear on a cold day?

▲ We can have a picnic on a sunny day.

We want to know how
to stay safe, too.

▲ People need to know about
bad weather to stay safe.

We can predict that the weather will be warm. What clothes will you wear? What will you do?

	What do the clouds look like?	Which way is the wind blowing?	What is the temperature?
Sunday		from the south	78 degrees Fahrenheit 26 degrees Celsius